CAREERS IN
INDUSTRIAL ENGINEERING

INDUSTRIAL ENGINEERS USE A COMBINATION of engineering skills and business acumen to help organizations run better. They consider factors such as location, supplies, inventory, technology, money, and the needs of workers to create systems that are more efficient, profitable, and safe. They strive to make products or provide services of the highest possible quality, while maintaining healthy and safe workplace environments. In the manufacturing arena, they design the workstations, automation, and robotics for systems all along the supply chain. They even design the entire workings of the factories. Within any industry, they can devise ways to do more with less.

The word "industrial" does not necessarily mean the work only applies to manufacturing. Although industrial engineers are found in nearly all manufacturing companies, the scope of their work is valuable in entertainment, shipping, healthcare, transportation, real estate development, and food service, to name a few. In recent years, fields like energy and IT (information technology) have become particularly reliant on the skills of industrial engineers.

Industrial engineering is one of the most versatile of the engineering disciplines, with many areas of specialization. It is practiced in all levels of an organization and can lead to many career choices, from data analyst to CEO. Daily tasks and project goals vary widely, depending on the job title, type of project, and employer. For example, industrial engineers made surgery easier for doctors by developing the system in which a nurse passes instruments to the surgeon. Other industrial engineers simplified a supply chain for UPS to make deliveries faster and easier to track. These are two very different projects that utilize the same basic engineering skills.

A bachelor's degree is required to become an industrial engineer. College degree programs in industrial engineering are very diverse, especially compared to other engineering disciplines. In general, industrial engineering majors learn to use engineering and scientific principles to design, manufacture, or improve systems that involve both goods and services. They are trained to take into account every conceivable variable, from budgets, to machine capabilities, to human imagination and error. They are taught how products are created, and how to improve the quality of those products at the lowest possible cost.

Of the 250,000 industrial engineers currently employed in the US, nearly 70 percent work in manufacturing, but there are many more opportunities outside of manufacturing for budding industrial engineers to consider. Some industrial engineers hold high-level positions in government agencies. Others apply their skills in organizations as diverse as banking, aeronautics, publishing, and entertainment. The outlook is good because industrial engineering skills are needed practically everywhere, and the demand is growing.

WHAT YOU CAN DO NOW

INDUSTRIAL ENGINEERING REQUIRES a strong educational foundation. Students interested in the field should start preparing early for admission into an engineering program at a college or university. Even as early as middle school, students can get ready by taking a variety of math and science classes. In high school, the focus should continue to be on math and science, but at a more challenging level. Ideally, students will take Advanced Placement (AP) college level courses like calculus, statistics, computer science, chemistry, and physics. English classes should also be taken seriously. Industrial engineers need to be able to read and write complex materials. Computer skills are a necessity, too, including at least some basic computer programming.

As a high school freshman, it is important to look into several college choices. Order their course catalogs and make sure your curriculum includes all the necessary classes to meet admission requirements.

Good grades are essential, but admissions officers will also be looking at what you have done outside of the classroom. Join school clubs that match your interests and aspirations. Take a leadership role to get a sense of how people work together. Participate in science fairs and any programs or projects that will expose you to engineering concepts. Engineering competitions offer a particularly fun way of doing that. FIRST (For the Inspiration and Recognition of Science and Technology) is just one of many organizations that sponsors all kinds of engineering challenges.

To get a good idea whether industrial engineering is truly the career for you, you should talk to the people who are doing this kind of work. Ask your school to arrange a day trip to a company where you can see engineers in their actual work environments. Better yet, arrange to job shadow a few industrial engineers in various industries in order to see the wide range of career opportunities that are possible.

The work can be challenging and the hours long, but most industrial engineers are happy with their career choice. Benefits include good working conditions, great pay, and opportunities for rapid advancement. If you are a natural problem-solver, enjoy math and science, and have strong people skills, take a look at industrial engineering.

HISTORY OF THE CAREER

THE CONCEPT OF INDUSTRIAL ENGINEERING was introduced in 1776 by Scottish philosopher and political economist Adam Smith. In his treatise, *The Wealth of Nations,* Smith wrote that division of labor was the key to economic growth. If workers were able to focus on specific tasks, rather than be required to do everything, overall efficiency would increase substantially. The idea influenced the technological innovators of the time to transform production systems from traditional start-to-finish manual operations into assembly lines where each worker focuses on one task in the production process.

As a profession, industrial engineering has its roots in the Industrial Revolution (1820-1870). Guiding the transition to mass production (the hallmark of the Industrial Revolution) was no easy task. In order to implement the division of labor concept, for example, someone had to determine how large operations would be broken down into many small components that would be divvied up among many workers. At the same time, new technologies, such as the flying shuttle and the spinning jenny, were introduced that changed manual tasks into mechanized operations. The steam engine, in particular, greatly accelerated the progress of manufacturing facilities. Other concepts soon came into play, including cost control systems to reduce waste and skills training for workers that were being redirected to production lines.

Math professor Charles Babbage had a major influence on the fledgling field of industrial engineering. Published in 1832, his book *On the Economy of Machinery and Manufacturers* reported his observations during visits to factories throughout England and the United States. Babbage detailed a number of different topics dealing with manufacturing. Some were mundane, such as how to determine the time required to perform a specific task. Others were more esoteric, such as the effect of learning on the generation of waste. The book was, in essence, the first textbook for industrial engineers.

Towards the end of the 19th century, developments in the industrial engineering field were more focused on how pay variations affected workers and by extension, productivity. Studies by Henry Towne and Fredrick Halsey, two members of the American Society of Mechanical Engineers (ASME), demonstrated that wage incentive plans could increase the productivity of workers without negatively affecting the cost of production. The plan proposed by Towne and Halsey further suggested that some of the gains be shared with the employees. It was a revolutionary idea then, but today, profit sharing is common.

After the turn of the century, the field of industrial engineering flourished. Henry Ford implemented various industrial engineering concepts, including the assembly line and incentive pay. He was able to reduce the time it took to build a car from more than 700 to 1.5 hours – a monumental success by anyone's standards. Yet there were many more aspects that needed to be explored and defined. Industrialists of the time came to realize that manufacturing is affected by many things – inflation, demand, public perception, seasonal fluctuations, national employment levels, advertising methods, inventory levels, and much more. There wasn't just one pioneer of industrial engineering contributing to the knowledge base, there were many.

Frank and Lillian Gilbreth first developed time and motion studies, a cornerstone of industrial engineering. The husband and wife team classified basic human motions into 18 basic elements ranging from noneffective to highly effective. These classifications, known as "therbligs," allowed analysts to design jobs without knowing precisely how much time was required to accomplish the work. They surmised that the time to complete an effective therblig could be shortened, while noneffective therbligs should be eliminated wherever possible. The Gilbreths' work led to the standardization of steelworkers' motions, which significantly increased output. It eventually led to a much broader field within industrial engineering known as human factors or ergonomics.

Frederick Taylor was another well-known pioneer in industrial engineering. His work involved matters such as organization of work, worker selection, and design of training programs. Trained as a mechanical engineer, Taylor believed in applying a high level of science to the pursuit of efficiency. For example, his books, *Shop Management* and *The Principles of Scientific Management*, laid out very precise methods for predicting manual tasks, improving work methods, developing work standards, and reducing the time required to carry out any task.

Mechanical engineer and management consultant, Henry Laurence Gantt, also believed in applying science to production and management issues. Gantt is best known for creating the Gantt chart in 1912. It is a cascading horizontal bar chart that makes it easy to visualize timelines for multiple tasks or events. The chart is a popular project management tool still used today for the scheduling of work.

Even as the profession was being refined from every angle, various institutions were developing training programs for industrial engineers. In 1909, Pennsylvania State University established the first department of industrial and manufacturing engineering in the US. Cornell University awarded the first doctoral degree in industrial engineering in 1933. The American Institute for Industrial Engineers was established in 1948. The association's primary purpose was to advance the technical and managerial excellence of industrial engineers. It also provided professional recognition for practicing engineers who previously had no real place in the hierarchy of a company.

Modern Practices

Throughout the 20th century, the industrial engineering field continued to grow. In the 1960s, the emphasis among industries around the world was on supply systems, inventory controls, transportation issues, and more efficient maintenance programs. In the 1970s, Japan obtained very high levels of quality and productivity by applying the management theories of Kaizen and Kanban. The results were so impressive that companies in the west adopted similar versions of "continuous improvement programs." In the 1990s, the world experienced the shift to globalization. As global industries sought to solidify their processes, the focus turned to supply-chain management, and customer-oriented business process design.

Future of the Career

What is the future for industrial engineers? Computers will certainly continue to make the job more efficient. More sophisticated analytical methods and advancing technologies make modeling complex production and business systems more of an everyday task. Of course, there will always be unpredictable factors that affect the outlook for industrial engineers. These professionals will always be in demand, not only in the United States, but also throughout the industrialized nations of the world.

WHERE YOU WILL WORK

THERE ARE ABOUT A QUARTER MILLION industrial engineers at work in the US. The large majority of these professionals – around 70 percent – work at manufacturing companies. Many have specific areas of specialization, such as assembly, raw-product processing, or administrative practices. The top five employers of industrial engineers include:

- Computer and electronic product manufacturing
- Machinery manufacturing
- Aerospace product and parts manufacturing
- Motor vehicle parts manufacturing
- Engineering services

This is a broad field that offers opportunities beyond manufacturing. Growing areas of employment for industrial engineers include:

- Hospitals and other healthcare operations
- Transportation
- Food processing
- Media and entertainment
- Banking
- Utilities

- Local, regional and national governments
- Aerospace
- Energy
- Hospitality

Overall, industrial engineers enjoy good working conditions. Work settings differ according to the type of projects or specific tasks. Work is usually performed either in offices or in the settings the engineers are trying to improve. For example, while trying to increase productivity in an industrial plant, an industrial engineer may spend considerable time on the factory floor, watching workers as they assemble parts. Then, after gathering sufficient data, they may work in an office analyzing the data on a computer. Industrial engineers also work in laboratories and on construction sites.

Travel is sometimes needed in order to observe processes and make assessments in various work settings. Those involved in global commerce make frequent trips abroad.

Work Schedules

Most industrial engineers work full time. The actual number of hours and how they are scheduled vary by industry and also by the type of projects being worked on. Hours can be long, but industrial engineers say this is outweighed by the high level of satisfaction derived from intellectually challenging work. Deadlines are common, but usually are not so tight as to require working late nights, weekends, or holidays.

THE WORK YOU WILL DO

INDUSTRIAL ENGINEERS TAKE WHAT OTHER ENGINEERS have done and make it better – and they do it in almost every industry. For example, in the entertainment field mechanical engineers design exciting new rides for Walt Disney World, while industrial engineers make guests happier by making the lines to get on those rides shorter. In the manufacturing industry, automotive engineers design the machinery and tooling needed to build cars for General Motors, while industrial engineers streamline the processes to make Chevrolets more affordable.

The goals of industrial engineers vary according to each project, but generally they strive to improve efficiency, raise productivity, reduce waste, lower costs, improve safety, ensure quality, and boost satisfaction.

To accomplish these goals, industrial engineers first study project requirements carefully. They analyze every detail, looking for flaws in existing processes. They then use mathematical models to develop new systems that would overcome those flaws. Industrial engineers in the manufacturing industry would typically follow these steps:

Review engineering specifications, production schedules, and factory process flows

Consult with vendors about purchases and seasonal fluctuations in supplies

Meet with management to determine manufacturing capabilities and available work force

Confer with workers about the status of projects and solicit their suggestions for improvements

Look for bottlenecks in supply chains and distribution systems

Develop management control systems to make budgeting more accurate

Implement quality control procedures to reduce the costs of

Unlike engineers in other specialties who may be focused on customer returns purely technological matters, the industrial engineer is involved with every aspect of an organization from personnel management to delivery methods to financial planning. At the start of each project, the challenge is to quickly become an expert not only in the processes of the particular industry, but also in the specific culture, problems, and challenges that the company faces.

New engineers typically work under the supervision of experienced senior industrial engineers. For many, the junior engineer's role is that of an assistant. The daily routine is mundane, with tasks restricted to collecting and organizing data, conducting low-level analysis, and cowriting reports with superiors. At first, raw data collected is fairly basic – number of employees, facility size and space, number of units being produced, and timings of shifts. As more knowledge and experience are obtained, engineers start doing more difficult tasks, such as time and motion studies using sophisticated software like Quetech. Eventually, they are assigned to a team where each member is responsible for very specific components within the overall project.

Career Options

Industrial engineers possess a broad range of knowledge and skills that can be applied to a variety of industries and positions. Job functions extend from managing McDonald's real estate, to planning guest flow at theme parks, to developing spaceport technology for NASA. There are many different job titles, each indicating the type of work being done. Some of the most common include:

Quality Engineers

These specialists work in quality control, sometimes referred to as quality assurance. In this role, they use scientific methods to measure, test, and ensure the quality and safety of products or services. By isolating each step of a business or manufacturing process, they can spot where changes could reduce production problems, mechanical malfunctions, or human errors. After implementing changes, the engineer continues to conduct audits to make sure the new process is standardized and low-cost. Quality engineers are vital to manufacturing, but they are in other industries as well. For example, they might be tasked with designing an improved admissions procedure at a hospital or develop a software protection program for spacecraft.

Project Manager

It is common for industrial engineers of all types to eventually become project managers because they possess an understanding of so many different aspects of business, and they tend to have good leadership and people skills. The job of a project manager is to plan and oversee the creation of products from beginning to end. That means organizing the human and material resources necessary to keep the project running smoothly while adhering to the planned budget and timeline. The project manager delegates responsibilities to team members and monitors the project's progress. It is also the project manager who is ultimately responsible for solving any problems that come up.

Healthcare Management Engineer

Industrial engineers play an integral role in the healthcare industry. Often known as "management engineers," they are specialists with a deep understanding of the healthcare industry. Their main job is to solve problems in the workplace, which is usually a hospital or other large medical facility. Typical projects might include determining optimal staffing levels or implementing new technology that reduces the burden of paperwork.

Lean Coordinator

The term "lean" in business refers to the reduction of waste so that (ideally) what is left are only value-adding processes and activities. Lean coordinators are found in all types of industries, but most commonly in manufacturing. These professionals are expert in identifying areas where costs could be reduced without sacrificing quality, safety, or satisfaction. Once a problem has been pinpointed, they devise a plan for improvement and lead a team to put the changes in place.

Supply Chain Solutions Engineer

These professionals find the best ways to arrange people, organizations, information, technology, and activities to transform raw components into finished products and deliver them to customers. This is an involved process that can be broken down into five stages: planning, development, manufacturing, logistics, and customer returns. As the job title suggests, the main focus is usually on planning a profitable and dependable supply of raw materials.

Cost Engineer

These professionals apply scientific principles to estimating the cost and control of a project. Engineering skills are used to determine the potential cost of a project and come up with accurate estimates for purchasing, scheduling, labor, packaging, and shipping. Because cost is a vital factor in nearly every business, there are many different types of cost engineering jobs.

Ergonomist

Also known as "human factors specialists," these industrial engineers are concerned with health and productivity in the work environment. When employees are comfortable, efficiency and productivity rises. Most ergonomists work for research and development divisions of manufacturing plants or private consulting firms. Manufacturers of office equipment and furniture hire ergonomists to research common health complaints and help develop products that maximize comfort and safety. Acting as consultants, these professionals tour business facilities and identify potential hazards, such as poor lighting, dirty equipment, and outdated technologies.

Sales Engineer

In many fields, the technologies involved are too complex for traditional sales representatives to understand and explain to customers. Industrial engineers with a flair for influencing potential buyers can fill the roll. Using a combination of technical expertise and communications skills, they can point out the benefits of a product and answer any technical questions a customer may have. These industrial engineers often work in software and hardware sales.

The Career Ladder

College degree programs in industrial engineering provide a good foundation for those entering the field, yet there is much more for graduates to learn about working in the real world. Large companies typically offer additional training to beginners, either through seminars or formal on-site classes. In the manufacturing industry, beginners are often required to work on the floor for a few months to experience firsthand the challenges faced in production.

For at least two years, young industrial engineers have limited responsibility and virtually no client contact. Gradually, they attain autonomy and develop plans, make decisions, and solve problems on their own. Eventually, junior industrial engineers become senior industrial engineers, and may advance to become technical specialists, such as quality engineers or facility planners.

By five years, industrial engineers have experienced a variety of problems and worked on teams and by themselves to solve them. Many meet with senior managers to discuss suggestions, improvements, and budgetary decisions. About 10 percent move out of the industrial engineering field into management positions by this stage. Those with people-managing skills may achieve the title senior industrial engineer and be in charge of managing others. Those in consulting firms lead teams instead of merely working on them. Hours increase, salaries increase, but more for those in consulting than for those in manufacturing.

Veterans of industrial engineering who have worked in the profession for 10 years or more have seen their salaries increase while their hours decreased. At this point, there are three general career tracks. Those who have stayed with the same employer typically move into managerial positions. Those who work for consulting firms are elevated to vice president level and have extensive client contact. Those who have chosen to move through a variety of companies are usually those who enjoy the industrial engineering process. They will likely continue to pursue the work they were trained to do and avoid the management route.

STORIES OF INDUSTRIAL ENGINEERS

I Am an Industrial and Systems Manufacturing Engineer

"I work in the automotive industry. My job is to evaluate and analyze problems that arise in the various manufacturing departments. Think of it as trouble shooting on a very big scale. Typically, people working in those departments know when something isn't working right, but are unable to pinpoint the issue or figure out what to do about it. However, those people are key to my successfully working it out. I direct them to collect necessary data and then rely on their expertise in their particular area to validate my proposed solutions. While I appear to be working solo, there is actually a full team helping me every step of the way.

There is a nice balance to this kind of work. About half my time is spent doing engineering tasks on a computer. The rest of the time I'm on the manufacturing floor, checking out specific real-world problems. There is also a balance between technical and creative abilities, which are applied equally. The needed technical skills are learned in school, but learning to apply those skills is a process that is learned through experience. I started out using a systematic problem solving approach. That works, but it is not likely to produce the most robust solutions. Lateral thinking and a good imagination are helpful. You also must be willing to learn new things and be open to input from others.

I love this work because it is intellectually challenging. I like being the hero who comes in and solves a problem that has perplexed everyone else. I also know that the future will bring even greater challenges as we try to keep up with advancing technologies. It will force us to be even more creative in our ideas.

I would advise anyone considering this career to get as much exposure as possible. Job shadowing and talking with people who are in the field are invaluable. There are a lot of possible career paths in industrial engineering. The more you look at, the better. Once you've made your choice, all you need (aside from a good education, of course) is the motivation to succeed."

I Am a Team Leader

"When I was in high school and thinking about my future, I told my counselor I wanted to be a math teacher. I was good at math, but didn't know what else I could do with that skill. The counselor handed me a book on potential fields to explore and bookmarked a few she thought might be a good fit. Two of them captured my interest, mechanical and industrial engineering. Since I was one of those kids who was always taking things apart to see how they tick, the idea of working with machines was appealing. Industrial engineering, however, goes beyond machines to integrate human factors and business systems. I was intrigued by what seemed like a more holistic approach.

Since making the decision to become an industrial engineer, I have worked in a variety of industries including public works, business development, marketing, construction, global distribution, and of course, manufacturing. That is one of the best things about this career. You can work at anything that interests you. Why get stuck in one type of job when you can experience so many different fields?

Industrial engineering has opened many doors for me. It has also opened my mind. I believe it has made me a better person, too. Through my work, I have explored life, come to understand people, learned to think outside of the box, invented, and created something out of nothing.

For someone who is results oriented, the process of making an idea or a concept into a reality that could change people's quality of life is immensely gratifying. Throughout my career there have been many challenges which have forced me to push myself. I have learned to embrace them all, considering each an opportunity for personal and professional growth. I encourage anyone interested in industrial engineering to learn more about it."

I Keep Roller Coasters Safe

"I've always loved the thrill of theme park rides, especially roller coasters. It has been my dream since I was a kid to work at a theme park, even long before I realized that engineers designed and built the rides. Today, I have been working at a destination resort that features some of the most exciting rides in the country.

My job is to ensure the safety of our guests at all times. It is always priority number one. I oversee the daily and seasonal inspections and maintenance. Every day the rides are each carefully tested before opening. Any malfunctions or repairs, however slight, are immediately taken care of by my engineering team. In the off-season, each attraction is completely disassembled. Every component is checked to make sure it's in perfect condition, and then the attraction is rebuilt. Finally, the ride is put through a full functional test to prove it is safe to operate.

What I like most about engineering is finding solutions to the problems that inevitably arise. For example, we had a simulator ride that needed a better seat belt mechanism. It was working, but in my judgment, it needed an upgrade. I spent nearly a year redesigning the mechanism, building prototypes, debugging it, and making sure it fit into existing seats. It was challenging, especially since I had never worked on anything like that. That is why industrial engineering is so interesting. My skill set can be applied to just about any situation. When I am able to rectify a problem, I feel a great sense of achievement. I am pleased to think that my work creates a safe environment for families coming to the resort for a great time.".

PERSONAL QUALITIES

ARE YOU SOMEONE WHO LOVES ORDER? Do you enjoy a challenging puzzle? Are you an innovative thinker who likes to think outside of the box? If so, industrial engineering might be a good career choice for you. Industrial engineers are inquisitive, possess leadership skills, and take initiative. The most successful professionals in this field also share the following common traits:

Creativity

It takes creativity and ingenuity to design new production processes in a variety of settings. Industrial engineers are constantly brainstorming new ways to reduce the use of material resources, time, and labor. Industrial engineering is a suitable career for anyone who enjoys problem solving and finding not just any solution, but the best one.

Critical Thinking Skills

Industrial engineers are analytical. Solving problems related to waste and inefficiency requires logic and reasoning. The most important skill needed for this job is a systematic problem solving approach. Industrial engineers analyze every detail of a process in order to identify the strengths and weaknesses of the possible solutions and alternative approaches. Attention to detail is also important. Without it, you may find yourself arriving at the wrong conclusions. This can be a challenge since the job often requires dealing with several issues at once, from aging technology, to workers' safety, to quality assurance.

Communications

Communications is one of the most valuable skills to have. Industrial engineers need to listen, explain, and teach. Even the best idea in the world will not get off the drawing board if it is not conveyed well. Successful industrial engineers are great communicators, able to convince others of the importance and urgency of their solutions. First, they listen. They listen to customers, vendors, managers, and technical staff in order to fully grasp the full scope of the problem. Then, within their own teams, they invite feedback and fresh ideas. Once a solution has been presented, an industrial engineer may need to teach technicians and production staff how to transition to a new way of doing things. Being able to quickly and clearly explain concepts is crucial to preventing loss of time and resources. Writing is also a significant part of the job. Industrial engineers write reports for management, prepare documentation for other engineers and scientists, and give presentations.

People Skills

Industrial engineering is not just about machines and processes – it is also about people. It is natural for company executives and employees to be resistant to change. Industrial engineers must be tactful in what they say and how they say it. Successful industrial engineers are pleasant, yet willing to stand firm by their recommendations.

A Knack for Numbers

The principles of calculus, trigonometry, statistics, and other advanced math topics are used every day in the life of an industrial engineer. Likewise, science and computer skills are important for analysis, design, and troubleshooting. If you lack math and technical skills, industrial engineering is probably not for you.

Motivation

The motivation to succeed is important when the going gets tough. It is a character trait that elevates an industrial engineer above the rest of the team when budgets are tight and deadlines are rapidly approaching. It is also necessary for long-term success in this career. Employers assume industrial engineers want to progress into managerial roles. Those who exhibit leadership skills, personal drive, and ambition are rewarded with early promotions.

ATTRACTIVE FEATURES

INDUSTRIAL ENGINEERING CAN BE AN EXCITING and rewarding career with both tangible and intangible benefits. The most obvious benefit is the exceptional pay. The average starting salary for new graduates is $55,000 a year. In some areas, salaries for entry-level jobs can go as high as $65,000. That is just a starting point. Salaries will continue to rise as your career progresses. It is common for experienced industrial engineers to earn more than $100,000 a year. Some also get bonuses and profit sharing on top of the base salary.

It is good to know that there will be a monetary payoff at the end of four grueling years of college. Yet most industrial engineers cite flexibility as the number one benefit this career has to offer. Industrial engineering is a very broad field, found in virtually every industry. Manufacturing, healthcare, technology, marketing – there are sure to be industrial engineers at work in nearly every company. This means no one needs to be stuck doing the same thing in the same industry throughout an entire career. Industrial engineering skills are transferrable to a great number of industries. For example, a new graduate might get started in manufacturing since that is one of the largest employers of industrial engineers. After a few years, manufacturing may become less interesting, and the idea of traveling the world seems exciting. No problem – there are a number of global commerce fields that feature frequent travel between countries.

Industrial engineers can also switch roles within an industry. In fact, career options are nearly limitless for industrial engineers. You can essentially create your own dream job by exploring and working in different jobs. For example, if you spend your days at a computer doing deep data analysis, but you really prefer to be working with people, a position in project management might be more satisfying.

The future looks bright. This is a stable field with constant demand for qualified professionals. The number of available jobs continues to grow because the range of services that industrial engineers are trained to provide is extensive and expanding. The career is nearly recession proof. Companies are always looking for ways to cut costs, improve processes, and gain a competitive edge. That is precisely where industrial engineers can help. Thus, from the business point of view, hiring an industrial engineer is not an additional burden and an expense, it is an investment.

There are plenty of advancement opportunities, with or without additional training. An experienced industrial engineer is a valuable commodity. It is not uncommon to be promoted after a few years to some level of management. In addition to an array of technical subjects, industrial engineering degree programs cover human factors, organizational management, and other business topics. That makes it an ideal career for someone interested in getting into management, even at the highest levels.

UNATTRACTIVE ASPECTS

THE FIRST CHALLENGE IS GETTING through college. The industrial engineering degree program is quite rigorous. It covers a wide range of difficult topics from robotics to economics to human efficiency, and everything in between. If you handily managed a load of AP courses in high school, you might be surprised when you do not breeze through college as well. The good news is you only have to buckle down for four years. A graduate degree, if you choose to pursue one, can wait until after your career is established.

Any job that involves problem solving is bound to create pressure. Industrial engineers routinely solve problems on a massive scale. Not all problems can be solved. Nor are all problems purely technical – though constantly changing technology is perhaps the most common challenge. Industrial engineers must deal with budget constraints, legal restrictions, management requirements, dynamic business conditions, and myriad environmental and social issues. Arriving at the most efficient solution is satisfying, but it is not always acceptable on all counts. Being forced to go with your second or third choice can be frustrating.

Industrial engineering is not known as a stressful occupation. However, some tension in the work place can build up due to looming deadlines, tight budgets, and people who do not understand how difficult it can be to solve complex problems. Expectations can be unreasonably high. People who are not engineers do not always understand what is possible and what is not. That includes the very same managers who are looking for solutions. Some of them may be demanding and unreasonable. Failure to meet expectations could mean a lost bonus or even unemployment. It may seem unfair, but that is the reality.

EDUCATION AND TRAINING

SOME CAMPUS AND ONLINE COLLEGES offer associate degrees in industrial engineering technology. However, because the work of industrial engineers is vital to the safety of so many people, a bachelor's degree is generally required for entry-level positions. At this level, there is too much hands-on training involved to make online education viable. A bachelor's degree typically takes four years to complete.

A bachelor's degree in industrial engineering is ideal, but there are many industrial engineers who have majored in mechanical engineering, electrical engineering, manufacturing engineering, or general engineering. Industrial engineering majors may earn the Bachelor of Science (BS) or Bachelor of Science and Engineering (BSE) degree. Not every college offers a degree program specifically in industrial engineering, but there are now accredited degree programs in this major at more than 70 schools nationwide.

When choosing a college to attend, you might want to consider the national ranking of a school. For both undergraduate and graduate programs, Georgia Institute of Technology in Atlanta has been a long time leader. Other top schools (in order of most recent ranking) include:

- University of Michigan, Ann Arbor
- Purdue University, West Lafayette
- University of California, Berkeley
- Virginia Tech, Blacksburg
- Stanford University, Palo Alto
- Northwestern University, Evanston
- University of Wisconsin, Madison

- Massachusetts Institute of Technology (MIT), Cambridge
- Pennsylvania State University, University Park
- Cornell University, Ithaca
- University of Illinois, Urbana-Champaign

Undergraduate Curriculum

Bachelor's degree programs include a mix of classroom lectures and laboratory practice.

The typical curriculum includes a broad array of foundational math and science courses such as calculus, differential equations, statistics, chemistry, physics, static and dynamic mechanics, materials science, computer science, electronics/circuits, and engineering design. Students can also choose elective subjects. Economics, finance, management, business strategy, and any of the social sciences are particularly valuable on the job.

Courses specific to industrial engineering include:
- Stochastic modeling
- Industrial cost control
- Organizational management
- Methods and work measurement
- Supply chain logistics
- Simulation methodology
- Manufacturing processes
- Automation
- Operations research

- Engineering economics
- Optimization techniques

Graduate Degrees

It is not necessary to get a master's degree, at least not right away. A bachelor's degree is adequate for most beginning positions. However, many industrial engineers pursue an MBA (Master of Business Administration) degree or a master's degree in a particular area of specialization once they are in the work force. In general, graduate programs in industrial engineering are primarily for those who want to enter academia or pursue executive management positions.

Graduate school can represent a significant investment of time and money. There are some cooperative education plans that take only five or six years to complete. They combine classroom study with practical work, providing the opportunity for students to gain experience and to finance part of their education.

The usual graduate degree for industrial engineers is the Master of Science (MS) or Master of Science and Engineering (MSE). The typical MS curriculum is very focused on industrial engineering subjects such as:

- Engineering economics
- Operations optimization techniques
- Facilities design and work-space design
- Supply chain management and logistics
- Time and motion study
- Corporate planning

- Human factors and ergonomics (safety engineering)
- Robotics and Computer-aided manufacturing
- Product Development
- Quality enhancement and control

Doctoral programs are available for those interested in conducting research. While studying topics such as integrated systems and variance analysis, students explore engineering theories, methods of conducting academic research, and advanced manufacturing systems and information systems. Coursework is followed by oral exams and research for the final dissertation.

Experience

Internships are part of every industrial engineering program. The real world experience gained as an intern is invaluable. In addition, many colleges and universities offer cooperative education programs in which students gain practical experience while completing their education. Employers consider production or manufacturing experience to be particularly useful, but they also look favorably on administrative experience in non-manufacturing industries, such as insurance, healthcare, or finance.

Licenses and Certifications

No certification is required for industrial engineers. However, since industrial engineering is a broad field encompassing many different industries, there are instances where certification is necessary for a particular specialty. Several certification programs are offered by the Institute of Industrial Engineers (IIE), including engineering management, industrial engineering professional skills, enterprise risk manager, and lean supply chain management. Each certificate program consists of professional seminars and a final examination.

A license is not required for entry-level jobs. Experienced industrial engineers, however, may need to acquire a Professional Engineering (PE) license. For example, only a licensed engineer may submit engineering plans to a public authority, oversee the work of other engineers, or provide services directly to the public. Licensing is issued by the states. Requirements vary somewhat, but generally include:

- A bachelor's degree from an ABET-accredited engineering program

- A passing score on the Fundamentals of Engineering (FE) exam

- Relevant work experience, typically at least 4 years

- A passing score on the Professional Engineering (PE) exam

Some states require continuing education in order to keep licenses current. Most states recognize licenses from other states.

EARNINGS

INDUSTRIAL ENGINEERS ENJOY EXCELLENT EARNINGS. Even new graduates do well, with an average starting salary of about $55,000 a year. That is higher than most college graduates entering the workforce can expect, and it only gets better from there. The median salary for industrial engineers is almost $85,000 a year. Many industrial engineers earn more than $100,000 a year, and earnings can go as high as $130,000. Salaries are often boosted by bonuses and profit sharing.

Industrial engineer salaries vary depending on experience, industry, and geographic location. Typical annual wages for the top industries employing these professionals are as follows:

- Oil and gas extraction $130,000

- Pipeline transportation $125,000

- Petroleum and petroleum products wholesalers $120,000

- Aerospace product and parts manufacturing $95,000

- Computer and electronic product manufacturing $90,000

- Architectural and engineering services $90,000

- Business management $85,000

- Machinery manufacturing $80,000

- Motor vehicle parts manufacturing $75,000

Do you live in Alaska, Wyoming, California, Oregon, or Washington? The average annual salary in those five states is more than $100,000. Alaska tops the list at $115,000. At the bottom of the pay scale are states like Wisconsin, Iowa, Indiana, Florida, and Arkansas. The average annual salaries in those states are around $75,000 – still not bad when you consider the reasonable cost of living.

Industrial engineers generally earn more the longer they are in the profession. Individuals can advance to higher earning positions such as quality engineers or facility planners, where they supervise a team of engineers and technicians. To do so often requires a master's degree.

Most industrial engineers receive full benefits, such as the usual paid vacation, sick leave, health and life insurance, and a retirement plan.

OPPORTUNITIES

EMPLOYMENT OF INDUSTRIAL ENGINEERS is projected to grow at a modest rate of five percent over the next 10 years. While it may not be the fastest growing engineering discipline, it is still experiencing a gap between the number of job openings and the number of qualified candidates to fill them. Firms are always seeking new ways to reduce costs and raise productivity. That is precisely the goal of every industrial engineer who is specifically trained to develop more processes that are efficient and reduce costs, delays, and waste. Because these engineers can accomplish this for many types of organizations, a wide variety of employers will continue to seek their services.

This occupation enjoys the distinct advantage of being exceptionally versatile. Unlike other types of engineers who are usually highly specialized, industrial engineers are employed in a wide range of industries. These include major manufacturing industries, consulting and engineering services, research and development firms, wholesale trade, and others. Look at the healthcare industry, for example. Healthcare is growing – along with associated costs. There are significant changes taking place in how healthcare is delivered. These factors are creating demand for industrial engineers in professional, scientific, and consulting services firms working within the healthcare industry.

Historically, employment of industrial engineers has been concentrated in manufacturing industries. Overall employment in manufacturing has been on the decline, which does not bode well for industrial engineers. However, like all industries, manufacturing firms are eager for anything that will help increase productivity and revitalize America's competitive edge in the global marketplace. It is the industrial engineer who finds ways to employ cost-saving technologies such as robotics, automation, and other complex processes.

On a worldwide scale, industrial engineers are becoming international ambassadors for American companies. National boundaries to business are fading. Industrial engineers who are fluent in foreign languages and customs are highly sought after. International travel for these professionals is becoming the norm, as companies expand and conduct more and more business in other countries. These same individuals are important to manufacturing industries that are considering relocating from overseas to domestic sites.

Likely retirements over the next decade will create more openings for aspiring industrial engineers. Industrial engineering is a field that happens to staff one of the largest proportions of older workers. In fact, 25 percent of currently employed industrial engineers are 55 years or older. Within industries such as manufacturing, the aging workforce is a significant threat to survival. Technically oriented recent graduates can find a home in manufacturing firms that are struggling to stay competitive.

There are plenty of opportunities for advancement in this field. Some industrial engineers return to school to obtain specialized training in order to move up the career ladder. For example, they may get additional training in a technical area that qualifies them to become facility planners or quality control engineers.

The best opportunities are for experienced industrial engineers with exceptional management skills and business acumen. Because their work is similar to that done in most management positions, many industrial engineers leave the occupation to become managers. The progression typically begins with supervising teams of engineers and technicians. The next step might be taking leading roles in various departments. Those with the strongest educational background in business can advance into general management.

GETTING STARTED

THE SEARCH FOR YOUR FIRST JOB should start long before graduation day. Throughout your college years, you should be gathering referrals and letters of recommendations from professors and supervisors you meet during summer jobs and internships. Join professional associations and attend meetings, conferences, and seminars. Keep track of the people you meet and add them to your list of referrals. Use that collection of contacts to build a network that can help you find job leads.

Internships are essential to a successful career launch. Industrial engineering is a broad field. To enhance your chances of getting hired, get as much experience as you can in as many areas as you can. Do not stop after one internship – participate in as many possible. If a company you intern for appeals to you, let them know. It is very common for an internship to progress into full-time employment.

Industrial engineers are often recruited on campus. Recruiters announce upcoming visits well in advance. Stop by your campus job placement office regularly to see when recruiters are expected to visit and sign up for interviews. The job placement office can also help you polish your résumé and help you improve your interviewing skills. When meeting recruiters, show real interest in the companies they represent. Do your homework. Research the companies so you can speak intelligently about how they would benefit from adding you to their team.

Search for jobs off campus. Start looking on the Internet. General job boards are not the best source of jobs for industrial engineers. Look instead for specialized websites such as EngineeringJobs.com and Indeed.com. You can post your résumé for free on these sites to attract the attention of recruiters and hiring managers. Check into offline recruitment agencies. Choose agencies that specialize in engineering jobs.

Learn all you can about the types of companies you would like to work for. Visit company websites and keep current with their projects. When you come across a company that interests you, network through your contacts to try and get the name of someone in that company. Simply knowing a name of a real person can help open doors.

ASSOCIATIONS

■ **Institute of Industrial Engineers**
http://www.iienet2.org/Default.aspx

■ **National Society of Professional Engineers**
https://www.nspe.org

■ **Association for Manufacturing Excellence (AME)**
http://www.ame.org

■ **The Association for Manufacturing Technology (AMT)**
http://www.amtonline.org

■ **FIRST (For the Inspiration and Recognition of Science and Technology**
http://www.firstinspires.org

■ **National Association of Industrial Technology (NAIT)**
http://www.nait.org

■ **Society of Manufacturing Engineers (SME)**
http://www.sme.org

■ **Board of Certified Safety Professionals**
http://www.bcsp.org

■ **Society of Women Engineers**

http://societyofwomenengineers.swe.org

■ National Society of Black Engineers
www.nsbe.org/home.aspx

■ Association of Iron and Steel Technology Foundation
http://www.aist.org/home.aspx

PERIODICALS

■ Journal of Industrial Engineering and Management
http://www.jiem.org/index.php/jiem

■ Plant Engineering Magazine
http://www.plantengineering.com/magazine.html

■ ISE Magazine
http://www.isemagazine.org

■ Engineering Magazine
advancedmanufacturing.org

CAREERS REPORTS
www.amazon.com/author/careers

CAREERS INTERNET DATABASE
www.careers-internet.org

Information
service@careers-internet.org

www.ingramcontent.com/pod-product-compliance
Lightning Source LLC
Chambersburg PA
CBHW061236180526
45170CB00003B/1316